BEASTLY BATTLES

STELLER'S SEA EAGLE VS. PTERANODON

BY CHARLES C. HOFER

CAPSTONE PRESS
a capstone imprint

Published by Capstone Press, an imprint of Capstone.
1710 Roe Crest Drive, North Mankato, Minnesota 56003
capstonepub.com

Library of Congress Cataloging-in-Publication Data
is available on the Library of Congress website.
ISBN: 9781669088714 (hardcover)
ISBN: 9781669088875 (paperback)
ISBN: 9781669088752 (ebook PDF)

Summary:
Watch out from above! Two fierce beasts are ready to meet in battle! The Steller's sea eagle faces off against Pteranodon. One is the largest and deadliest eagle living today. The other is an ancient giant that soared through the skies during the time of dinosaurs. Who will claim victory?

Editorial Credits:
Editor: Peter Mavrikis; Designer: Bobbie Nuytten; Media Researcher: Svetlana Zhurkin; Production Specialist: Whitney Schaefer

Image Credits:
Capstone: Jon Hughes, cover (bottom), 5 (bottom), 7, 11, 15, 18, 19, 22, 23, 27 (bottom), 29; Dreamstime: Andrey Gudkov, 16; Getty Images: 500px/Jiri Hrebicek, 21, 500px/Jose Gieskes, 5 (top); Shutterstock: Andrey Gudkov, 24, 25, 27 (top), nutsiam, 9, Petr Simon, 6, Russ Jenkins, cover (top), Sergey Uryadnikov, 13, 17, Yoshiaki Katou, 20, Zaid Bin Yaseen, 28

Printed and bound in China. PO 6098

TABLE OF CONTENTS

Words in **bold** are in the glossary.

AIR COMBAT

It's a fight between two fierce fliers. Both are deadly **predators**. One ruled the sky during the time of the dinosaurs. The other is one of the largest birds living today.

Who would win the battle between these powerful beasts?

STELLER'S SEA EAGLE

PTERANODON

MEET THE FLIERS

The Steller's sea eagle lives along the **coasts** of eastern Asia. Its home is very **remote** and cold.

Pteranodon was one of the largest animals to ever fly. It lived millions of years ago, mainly in North America. Back then, much of the **continent** was covered by a sea.

STONE COLD HUNTER

The Steller's sea eagle is a **raptor**. These birds are predators. They hunt other animals for food. Other raptors include hawks and falcons.

This sea eagle has several deadly weapons. It uses its beak and claws to catch food and fight off threats.

WINGED REPTILE

Pteranodon lived alongside many deadly dinosaurs. However, Pteranodon was actually not a dinosaur! It was a flying **reptile**.

Pteranodon went **extinct** about 66 million years ago. Because of its size, Pteranodon ruled the skies during the time of dinosaurs.

SOARING GIANT

The Steller's sea eagle is a massive bird. Its **wingspan** can be over 8 feet (2.4 meters).

The Steller's sea eagle is covered in thick feathers. This helps it stay warm in the cold air.

Like all birds, the Steller's sea eagle has **hollow** bones. This makes the birds light, so they can fly. Even with hollow bones this eagle still weighs up to 20 pounds (9 kilograms).

RULERS OF THE SKY

Pteranodon also had hollow bones. But it did not have feathers. Instead, its wings were large flaps of skin.

Pteranodon was much larger than the Steller's sea eagle. Its wingspan could be more than 20 feet (6 m) across. This giant flying reptile could weigh up to 25 pounds (11 kg).

DEATH FROM ABOVE

The Steller's sea eagle has a giant beak. It can be up to 4.5 inches (11.4 centimeters) long. The sea eagle uses its beak to tear apart its food.

The Steller's sea eagle eats fish, crabs, and other birds. It will swoop in and snatch its food from the water.

TOOTHLESS TERROR

Pteranodon was an odd creature. It had a huge head with a long **crest**.

Pteranodon also had a long beak. Much like modern birds, Pteranodon did not have teeth. Since it lived near water, Pteranodon most likely ate a lot of fish.

SLASH AND CLAW

The Steller's sea eagle has deadly claws. These are called **talons**. The Steller's sea eagle uses its talons to catch and hold its prey.

Two Steller's sea eagles fighting

The mighty talons can be used as a defense too. Eagle talons are powerful and razor-sharp!

LAND BIRD

Pteranodon most likely walked on all four legs. But scientists believe it spent most of its time flying over the water looking for food.

Its large wings allowed it to soar long distances without getting tired.

EAGLE-EYED

The Steller's sea eagle has excellent vision. This helps it see prey from far away. No wonder good vision is called "eagle-eyed."

Steller's sea eagles have **binocular** vision. This lets them swoop in and catch a moving fish.

FLYING INTO BATTLE

Look, up in the sky! The Steller's sea eagle is hunting for prey. But Pteranodon is hungry too.

The two fantastic fliers meet in the air. The Steller's sea eagle snaps its powerful beak. Pteranodon flaps its massive wings. They're ready to fight!

WHO'S THE WINNER?

The Steller's sea eagle is one of the fiercest birds living today. Pteranodon soared through the skies long ago. Who would win this battle for the sky?

	Steller's Sea Eagle	Pteranodon
HABITAT	coasts of eastern Asia	the sea that once covered North America, as well as Europe and Asia
WINGSPAN	8 feet (2.4 m)	20 feet (6 m)
WEIGHT	up to 20 pounds (9 kg)	up to 25 pounds (11 kg)
WEAPONS	• sharp talons • large beak • excellent vision	• massive size • powerful beak

GLOSSARY

binocular (bi-NOCK-u-lar)—using both eyes with overlapping fields of view

coast (KOHST)—part of land near a sea or ocean

continent (KON-ti-nent)—a large land mass

crest (KREST)—a tuft of feather or skin on the head of an animal

extinct (ex-TINKT)—no longer living

hollow (HALL-oh)—having an empty space inside

predator (PRED-uh-tur)—an animal that hunts other animals for food

raptor (RAP-tor)—a bird of prey

remote (ree-MOTE)—a distant place

reptile (REP-tile)—cold-blooded animal with scaly skin that lays eggs

talon (TA-lon)—a sharp claw

wingspan (WING-span)—the distance between the tip of one wing to the tip of the other

READ MORE

Gish, Melissa. *Eagles*. Mankato, MN: Creative Company, 2020.

Markle, Sandra. *On the Hunt with Eagles*. Minneapolis: Lerner Publications, 2023.

Yang, Yang. *The Secrets of Pterosaurs*. Dallas: Brown Books Kids, 2021.

INTERNET SITES

American Museum of Natural History
amnh.org/exhibitions/pterosaurs-flight-in-the-age-of-dinosaurs/what-is-a-pterosaur

San Diego Zoo Wildlife Alliance
animals.sandiegozoo.org/animals/stellers-sea-eagle

INDEX

ABOUT THE AUTHOR

Charles C. Hofer is a writer and biologist living in New Mexico, in the land where Pteranodon once roamed.